D1826509

Laurance Reed

An ocean
of waste

Laurance Reed

An ocean of waste

Some proposals for clearing the seas around Britain

Conservative Political Centre
on behalf of
THE BOW GROUP

Bow Group pamphlets do not purport in any sense to represent an official Party view, nor even the collective opinion of the Group, but are published as containing facts and opinions which merit consideration by the Conservative Party and by a wider public.

CPC No. 500

Published by the Conservative Political Centre, 32 Smith Square, London SW1 and printed by Oliver Burridge & Co. Ltd., Crawley, Sussex.

First published January 1972

ISBN 0 85070 496 0

Contents

The author

LAURANCE REED is the Member of Parliament for Bolton East.

He is Joint Secretary of the Parliamentary and Scientific Committee, secretary of the All Party Committee on Pollution of the Environment and a member of the Select Committee on Science and Technology.

Before entering Parliament he wrote and lectured on ocean development. He retains an interest in this field as a director of the Association Européenne Océanique and as a member of the Society for Underwater Technology.

1. Introduction

THE OCEANS' CAPACITY to degrade and dilute waste products is enormous and to exploit them as a dumping ground for trade and domestic refuse is perfectly legitimate. But the self-purifying powers of the oceans are not unlimited and, unless some control is exercised over the nature and volume of wastes discharged, serious and lasting damage may be done to other uses and resources of the sea.

Coastal waters receive the larger part of marine disposals. These shallow, confined areas afford less dilution than the open sea as currents in the mixing process tend to foster the inshore concentration of pollutants. At the same time the coastal belt is the most used, the most productive and the most valuable of all the regions in the oceans. Any damage done to this zone would certainly have far-reaching consequences.

Extensive areas in the Great Lakes, the Caspian and Baltic Seas have already been converted into lifeless deserts by indiscriminate dumping, and a number of isolated basins adjoining industrialised countries which face the open sea also show signs of deterioration. DDT has been observed in places as far away as Antarctica and Thor Heyerdahls' accounts of oil pollution in mid-Atlantic are well known.

These reports cover only the surface layers of the sea. More striking testimony of the extent of marine pollution has been given by Jacques Cousteau. This man has spent a life-time exploring underwater and his observations are valid down to a depth of at least two hundred fathoms. Last year he told the Council of Europe that the intensity of life in all the seas of the world had declined by between 30 and 50 per cent during the past twenty years.

In the case of commercial fisheries it appears that the depletion of stocks is due as much to overfishing as to any other cause but Cousteau included in his estimate fixed fauna, vegetation, plankton, coral and other forms of marine life whose decline can only be attributed to the accumulative effects of pollution.

In the marginal seas around Britain there is no sign that waste disposal has adversely affected marine life on any scale. Where contamination persists it remains highly localised and on the few occasions when pollution has become widespread its effects have been shortlived. On the other hand anti-pollution laws on land increase the demand for sea disposals, and the shallow, semi-enclosed character of our offshore areas sets a limit on the type and volume of wastes that can be safely discharged there.

Britain, in common with all coastal states, has full jurisdiction inside territorial waters and she can regulate dumping by her own vessels anywhere at sea. Even so no comprehensive system of national controls has been established and there are serious shortcomings in the machinery for applying such rules as do exist. In an island as heavily industrialised as ours it would perhaps be unreasonable to expect coastal waters to be kept completely clean but it is not unreasonable to expect that waste disposals will be regulated so as to avoid needless damage.

In this field international cooperation is obviously essential. The specialised agencies of the United Nations have been active for some time but the need for further measures of prevention and control has been widely recognised. There is a growing conviction that urgent action is required and pollution of the sea has been placed high on the agenda of the Stockholm Conference on the Human Environment, 1972.

Stockholm will be a useful forum for discussing some of the legal problems involved in prevention and control. The enlargement of the European Community should also improve the climate for regional cooperation in this sphere. Nevertheless it would be unwise to put too much faith in an early international solution. Treaties and conventions customarily take three to four years to negotiate, and five to six years to ratify. To this a further decade can be added before any agreement would be properly enforced.

There is no reason, though, why our own Government should not act unilaterally to regulate or prohibit pollution originating from sources under national control. Britain really carries very little weight in protesting against the activities of other states when she herself fails to apply statutory controls. New legislation would strengthen our hand by setting a standard for other countries to live up to. The detailed rules and procedures to be adopted are best left to the expert. What Parliament can do is to provide the legal and administrative framework which will ensure their effective enforcement.

2. Major sources of pollution

THERE IS NO REAL EVIDENCE that marine pollution has seriously harmed marine life in British coastal waters. Where contamination is persistent it remains highly localised and on the few occasions when it has become widespread its effects have been shortlived. But our knowledge of what is happening to pollutants deposited in the sea and what long-term effects they are producing is very rudimentary, and caution must be the watchword.

It is known that contaminants which reach the sea in harmless dilutions can be accumulated by biological processes to the point where toxic levels are attained. Other substances, harmless in themselves, can be converted into lethal agents by interaction with seawater or with other pollutants which may be present. It also seems possible that although pollutants in our nearshore environment may not be present in sufficient concentrations to harm marine life directly, they may increase the chances that sea animals which are under stress from some other cause will die rather than recover.

This was the conclusion reached by the team investigating the deaths of over 12,000 seabirds, mostly guillemots, which were washed up on the shores of the Irish Sea and its approaches in the autumn of 1969. All kinds of culprits were named as responsible at the time but after some detailed detective work, the team found that no single factor, natural or artificial, could be regarded with certainty as the main cause of the incident, and that the most likely explanation of the mystery was that several factors, acting in combination, were responsible. If this conclusion is correct it is not re-assuring for by shifting the balance between life and death, pollution puts whole species of marine animals at risk.

There are a variety of routes by which pollutants can enter the marine environment but they all get there either by accident or design. The volume of contaminants from all sources is increasing and the shallow, semi-enclosed character of our offshore areas will aggravate the problems they create.

Marine disposal

Large volumes of waste are quite deliberately disposed of in the sea. They are discharged via rivers and from outfalls along the coast, or they are taken out to sea and dumped. In each case marine disposal is intended and therefore the source of pollution is capable of being controlled.

Coastal discharges

Rivers carry to the oceans sediments leached from the land as part of their natural load. In many industrial countries this geological rate of disposition is exceeded by the volume of human wastes. There are no figures which show the total quantity of man-made material deposited by our rivers but a large international highway like the Rhine pours over 60m. tons of dissolved solids into the North Sea every year. The quantity of waste products introduced from other sources is only a fraction of that contributed by rivers. Clearly no clean seas programme can be contemplated before pollution from this source has been checked.

Wastes discharged to the sea via rivers, however, are usually treated, whereas wastes deposited directly into marine waters normally are not. In England and Wales the sewage refuse of 6m. people living in coastal towns is piped into the sea.

Coastal discharges of industrial wastes

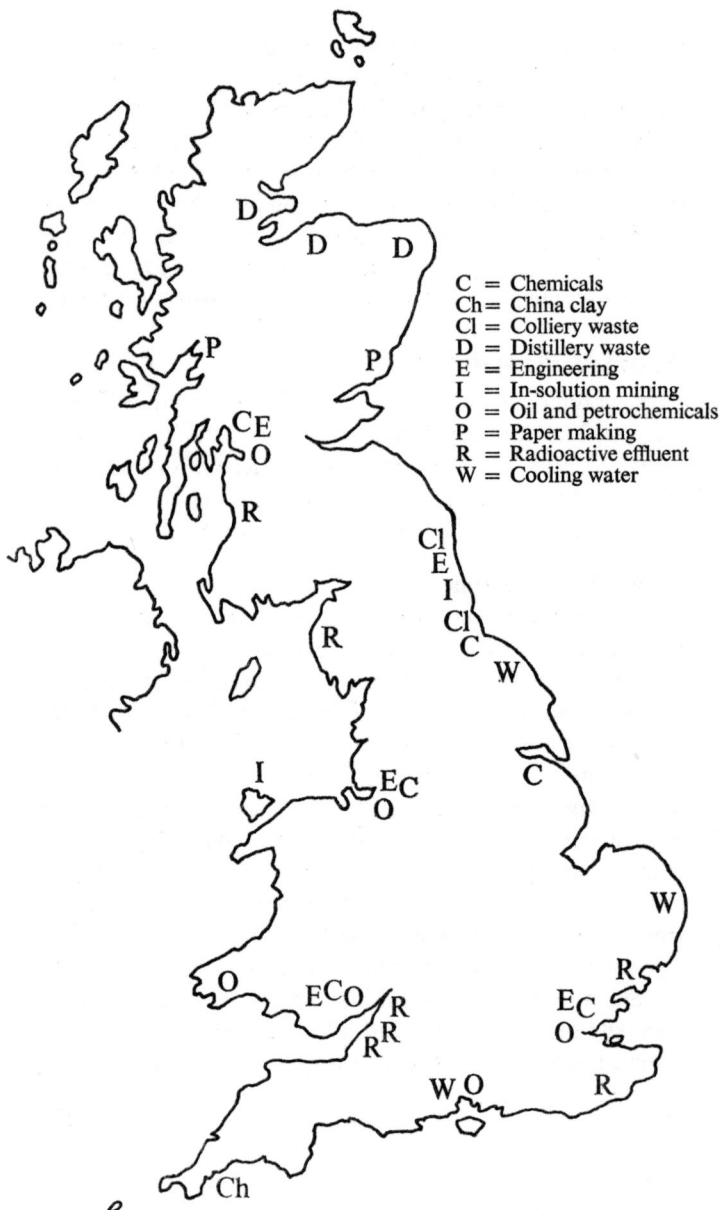

C = Chemicals
Ch = China clay
Cl = Colliery waste
D = Distillery waste
E = Engineering
I = In-solution mining
O = Oil and petrochemicals
P = Paper making
R = Radioactive effluent
W = Cooling water

Only twenty-nine of the one hundred and ninety-two local authorities involved employ full treatment works.

The most badly polluted areas are enclosed bodies of water situated near major conurbations like the Mersey, the Tees and the Tyne. The tidal Tyne, for example, carries to the sea 37m. gallons of sewage a day. Other heavily polluted estuaries in England are the Wear, Humber, Usk and Ribble. On the Welsh side of the Bristol Channel there are ninety outfalls discharging the sewage of almost half the population of Wales. Seven of the discharges are fully treated and nine partially treated. In Scotland the domestic waste of the capital is pumped into the Firth of Forth without any treatment whatsoever.

A river will carry its load some way out to sea but the majority of pipelines installed along the coast run out no further than low water mark. Longer sea outfalls however are being built or planned. In the Wirral, local authorities are constructing a three-mile long outfall to pipe 6m. gallons of untreated sewage a day into Liverpool Bay, and in the South the Greater London Council has considered a scheme for pumping sludge into Joss Bay, Kent via a pipeline network passing through Woolwich. Three miles out the line would divide to give two outlets about one and a half miles apart and nine miles offshore.

Industrial effluent and waste of all kinds are also discharged directly into coastal waters. Oil refineries, steel making, aluminium smelting, chemical processing, and other industries which find a locational advantage on the coast are the major polluters. Power stations, too, return huge quantities of waste heat to the sea. The Hunterston Generating Station on the Ayrshire coast discharges 20m. gallons of warmed cooling water an hour with a temperature 10°C above that of the surrounding seawater. Where the station is nuclear, radioactivity constitutes an additional hazard; especially in areas like the Bristol Channel where no less than six atomic reactors have been built or planned. The nuclear fuel processing plant at Windscale in Cumberland pumps radioactive effluent into the Irish Sea less than a mile and a half from shore.

Mining operations near the sea are another source of pollution. About six miles of the Durham coast have been hideously corrupted by millions of tons of colliery waste dumped on the foreshore. Tipping has also disfigured stretches of the coast in Northumberland, Cumberland and Flintshire. In Yorkshire, Cleveland Potash has permission to discharge 170,000 tons of potash mining waste annually into the sea near Staithes. St Austell and Mevagissey Bays in Cornwall are visibly polluted by china clay wastes deposited by rivers from workings upstream. About three-fifths of the seafloor in the region is affected. There is a plan now to pipe the waste overland and discharge it from a half-mile tunnel under the sea.

Industrial and commercial development covers about one hundred and sixty miles of coastline in England and Wales and obviously the volume of waste discharged will increase as the coast comes under more intensive use. Offshore airports, oil terminals, marinas, trunk sewers, tidal power stations and water retention barrages, in-solution mining and desalination plants will add significantly to the pollution load in nearshore waters. One in-solution mining plant at Hartlepool, which extracts magnesium from seawater, discharges 35 tons of suspended solids daily. The recently approved desalting project at Ipswich, designed to produce a million gallons of freshwater a day, will return to the estuary about 1·6m. tons of brine effluent annually with a salt concentration twice that of the water abstracted.

14

Ocean dumping

High-level radioactive waste in Britain is stored on land but in addition to the small quantities of less active effluent released into coastal waters some wastes are consigned to the deep beyond continental shelf areas. The dumping is usually performed by merchant ships on routine journeys and is carried out in accordance with internationally agreed rules. The materials are set in concrete so that any seepage would be very slow. The Atomic Energy Authority has considerable experience of deep sea dumping practices and offers industry advice about the disposal of difficult wastes on a repayment basis.

The Royal Navy is also in the deep sea disposal business. After the last war more than 200,000 tons of nerve and mustard gas stored in metal containers were dumped in the North Atlantic some two hundred miles off the west coast of Scotland. Sea disposal of surplus stores has recently been discontinued but the Navy still dumps ammunition in various deeps within the regions of the continental shelf. Areas listed as approved dumping grounds and marked on Admiralty charts are: the Hurd Deep, St Catherine's Deep, the Plymouth Spoil Ground (all in the English Channel); Milford Haven (Bristol Channel); May Island (Firth of Forth); and Beaufort's Dyke (North Channel).

Also marked on charts are spoil grounds used for the dumping of construction and demolition debris, and a variety of trade and municipal wastes. Purpose-built vessels are often used and at present about one-fifth of the sewage sludge produced by inland municipal works is disposed of in this way. The demand for sea disposal services is growing. Several firms specialise in taking wastes to sea and some local councils are developing facilities for small quantities of containerized wastes.

Approved dumping grounds in offshore areas

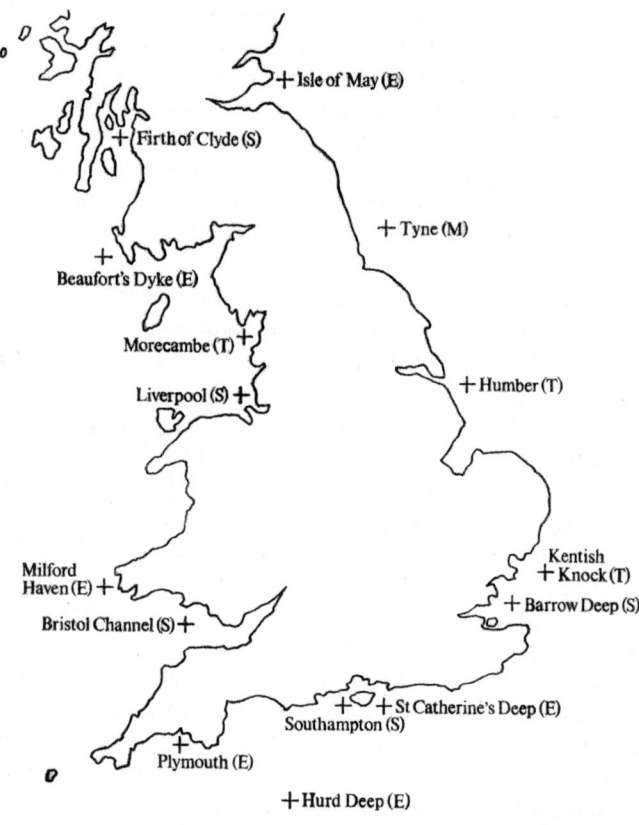

Isle of May (E)

Firth of Clyde (S)

Tyne (M)

Beaufort's Dyke (E)

Morecambe (T)

Humber (T)

Liverpool (S)

Kentish Knock (T)

Barrow Deep (S)

Milford Haven (E)

Bristol Channel (S)

St Catherine's Deep (E)

Southampton (S)

Plymouth (E)

Hurd Deep (E)

E = Explosives
M = Mining wastes
S = Sewage sludge
T = Trade wastes

The outer Thames Estuary has been in use since 1887 and currently 5m. tons of sewage sludge are dumped there each year. The Firth of Clyde has been used as a dumping ground by the Glasgow Corporation since the beginning of the century. Neighbouring authorities now exploit the same site and altogether about a million tons of sludge are dumped annually.

In the Irish Sea there are two spoil grounds – Liverpool Bay and Morecambe Bay. Both are about ten miles from land and have a water depth of fifteen fathoms. Approximately 2000 tons of industrial waste and 7000 tons of sewage sludge are dumped each week. The trade wastes come from customers like ICI and Fisons, and the sewage from the Manchester and Salford Corporations. Local authorities in south-east Lancashire are to build a trunk pipeline to carry their sludge direct to Liverpool for dumping in the bay.

Two new sites have recently been approved as dumping grounds in British coastal waters. One is in the southern North Sea and the other is in the Severn Estuary. The former is located near the Kentish Knock – a sandbank lying fifteen miles north-east of the North Foreland. It is expected that 400,000 tons of industrial waste will be dumped annually. Some of it will be coming from ports on the Continent.

Germany and other countries along the Rhine are currently shipping half a million tons of chemical waste down river for dumping in the North Sea. This represents a fourfold increase over the quantity dumped in 1968. It seems that part of the load is being shed in waters overlying UK continental shelf areas. These highly toxic products are encased in sealed containers. Some have been washed up on the coast and others have been found to be leaking when dredged up by fishing nets. Last summer a Norwegian research vessel sailed through a thick seventy-mile long bank of dead and dying

17

fish in an area of the North Sea where mackerel spawn. All the fish examined contained chlorinated hydrocarbons of the kind used in the plastics industry.

Large quantities of crude oil are also dumped in the sea by tankers when flushing out their oil tanks or pumping out contaminated bilge water. This practice is often referred to as 'operational pollution'. During a regular degassing operation at sea a 30,000 ton tanker would release over 100 tons of oil residues and it is estimated that at least 600,000 tons of oil a year find their way into the sea from this source.

Ocean spillage

Many noxious and poisonous substances are introduced into the marine environment by accident. Some of them are spilled by ships and some released by operations on the seabed. While steps can be taken to contain the number of incidents, involuntary pollution of this kind is, by definition, uncontrollable.

Hazardous cargoes

Possibly as many as 6m. tons of crude oil are lost to the sea annually. A great deal of this comes from land-based sources but at least 1m. tons gets spilled accidentally when tankers are being loaded or offloaded, or as a result of ship collisions and strandings. Since 25 per cent of the world's oil traffic is shipped through the Channel a large part of the spillage inevitably ends up on our coasts. The Coastguard Service recorded two hundred and seventy-two reports of oil at sea or on beaches during 1969, and three hundred and thirty-seven in 1970. In a year without a major accident about 50,000 tons of oil contaminate our beaches and inshore waters.

Crude oil is only one of a number of hazardous cargoes

transported in the shipping lanes around Britain. Others in the noxious freight category are: radioactive substances, nerve gas, ammonia, sulphuric acid, molten sulphur, molten phosphorous and liquified methane. Chemical and gas carriers may be smaller than oil tankers but their cargoes are infinitely more dangerous. An accident involving any one of these substances could assume catastrophic proportions.

It is impossible to say how far pollution from this source will increase but it seems certain that it will not decrease. The seaways around our shores are the busiest in the world. Three hundred thousand ships a year pass through the Dover Strait and in the summer months one hundred ferry sailings a day run across the mainstream of traffic. Already one in ten of all shipping accidents occur in the English Channel or its approaches, and no less than half the collisions between Dover and the Elbe.

In theory an increase in the size of ships should reduce the number of vessels, the density of traffic and hence the accident rate. But because bulk carriers and monster tankers lack manoeuvrability and have very deep draughts they tend to reduce the amount of sea-room available and increase the likelihood that ships will collide with one another. Super-tankers also run the risk of colliding with the seafloor.

The Torrey Canyon drew 45 feet. At 120,000 dwt she was among the thirteen largest vessels afloat in 1967. Today there are ninety-eight tankers of more than 100,000 tons in existence and at least six of these draw over 80 feet – twice the draught of the old Queens which were the largest ships in service in the 1950s. And we now face the prospect of the million-ton tanker a third of a mile in length and drawing between 100 and 120 feet.

A Japanese tanker once struck the bottom in the Malacca Strait but escaped undamaged. The next ship may not be so lucky. Alternative routes can add several days to a voyage and since time is money, there is considerable pressure on ship's masters to operate in ever more marginal conditions when navigating. It is believed that some masters accept a below hull clearance of only three to four feet in the Channel and even less when entering or leaving port.

This can only be considered reckless when one knows that 40 per cent of the seas near our coast have not been re-surveyed since they were first charted between 1840 and 1870 with lead and line. Even in the most up-to-date surveys soundings are seldom accurate beyond 60 feet because of the limits in accuracy with which the depth of the sea can be measured. A ship's list and pitch will lessen the amount of clearance available and the margin for error is further reduced by fluctuations which occur in predicted tide heights under abnormal meteorological conditions, as well as by changes in the level of the seafloor itself caused by the migration of sand waves over it. These sand waves can have amplitudes of up to 30 feet.

The changeable sea areas around Britain are cluttered with the wrecks of two world wars. There are about seven thousand ships at the bottom of the North Sea. Of these over four thousand are marked on Admiralty charts but for some only a 'position approximate' is recorded. During recent surveys in the Thames Estuary, North Sea and Dover Strait about twenty previously uncharted wrecks were discovered. These appear to be old wrecks long since covered by sand but now becoming uncovered as the bottom topography alters. Conversely, quite a few known wrecks could no longer be found and have presumably become covered.

Seabed operations

The continental shelf is a recognised source of minerals, especially hydrocarbons. At present 17 per cent of the world's oil supply comes from marine fields and by 1980 this could reach 40 per cent. Our own shelf areas meet 70 per cent of the country's natural gas needs and the reserves of crude oil discovered to date in the national sector of the North Sea are sufficient to supply three-quarters of our requirements at current rates of consumption. Other structures which show oil bearing potential have been located in the Minches, the Western Approaches and Rockall Bank.

The only hard mineral exploited at present is coal. Mining from above the shelf is impracticable and shafts are driven under the sea from mines on land. Seventeen collieries obtain all or part of their coal from this source, representing 9 per cent of the national output. Four mines have shafts extending beyond three miles. The longest, at Haigh in Cumberland, is just four miles long.

Britain also obtains 10 per cent of her sand and gravel from the seabed. Large deposits are exploited off the Lincolnshire coast, in the southern North Sea and in Liverpool Bay. Not much is known about other minerals in the national share of the continental shelf but alluvial tin has been located in the sea off Cornwall and lime deposits, useful as a fertilizer and in the manufacture of cement, have been found in the Minches. The seafloor beneath the North Sea is also thought to hold huge reserves of potash. Deposits lying along the Yorkshire coast are worked from the land.

It is impossible to assess the pollution threat from these activities but it must be recognised that some damage to the marine environment is unavoidable when exploiting the seabed. Dredging operations, for example, produce a suspended sediment problem. Extensive underwater landslips could also

Mineral exploitation of continental shelf

A = Aggregates
C = Coal
G = Gas fields
O = Oil finds
P = Potash
T = Tin

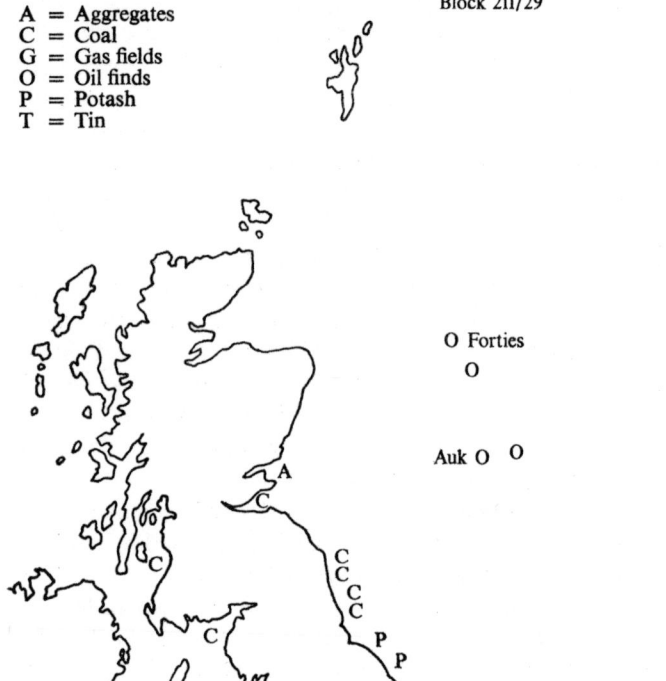

be caused which might disturb the vital upwelling processes that carry nutrients to the surface and enrich coastal waters.

Mining from beneath the shelf could cause pollution if the shafts were to be breached. One of the old tin mines in Cornwall was flooded in this way when a ship ran aground near the coast. Tin is not toxic even in high concentrations but marine life would be adversely affected if the workings were for metals of high toxicity like copper and zinc.

A submarine oil well 'blow-out' is less of a danger than is commonly supposed. Since 1954 more than eight thousand wells have been drilled off the US coast. Only two ever got out of control and caused widespread pollution. Nevertheless, when disaster does strike the results can be spectacular. The celebrated 'blow-out' in the Santa Barbara Channel smeared thirty miles of beach and spewed up enough oil to blacken eight hundred square miles of ocean. It took three months to plug the leak effectively and there is still some seepage to this day.

Underwater storage tanks and pipelines are a more serious problem. The technology of subsea pipeline laying is far from perfect and owing to the difficulties of constructing a line across the deep trough adjoining the Norwegian coast, Phillips Petroleum are to build a seabed storage tank for use at their Ekofisk Field. It will have a storage capacity of 35m. gallons and stand fifty feet above sea level. If a ship holed this tank there would be an appalling mess. Ships can also plough into oil rigs and in an area as congested as the North Sea this must be considered more of a probability than a possibility.

The heavy scouring action of bottom waves in the shallow North Sea tends to remove the overburden from completed pipelines. Once a section of pipe is uncovered resonance in the unsupported parts produces a snake-like motion which soon uncovers more. Exposed lengths are then apt to swing to and fro with the tides. In this state they are very vulnerable to ship's anchors and fishermen's nets. Pipelines underwater are an attraction to fish, and fishermen are tempted to trawl up and down the line despite the obvious risks.

The BP pipeline from West Sole Field to Easington now lies partly exposed on the seabed though it was originally buried. Divers inspecting the line have found dozens of trawl boards lodged against it. Fortunately methane is not a pollution risk as the gas escapes through the water column and is dissipated in the atmosphere. But a ruptured pipe can explode and shipping in the vicinity is at risk. In this way natural gas could be the indirect cause of pollution at sea. With oil in the pipeline the pollution threat would be immediate.

Neptune 7 at work near the Forties oil fields
(*Reproduced by kind permission of Shell*).

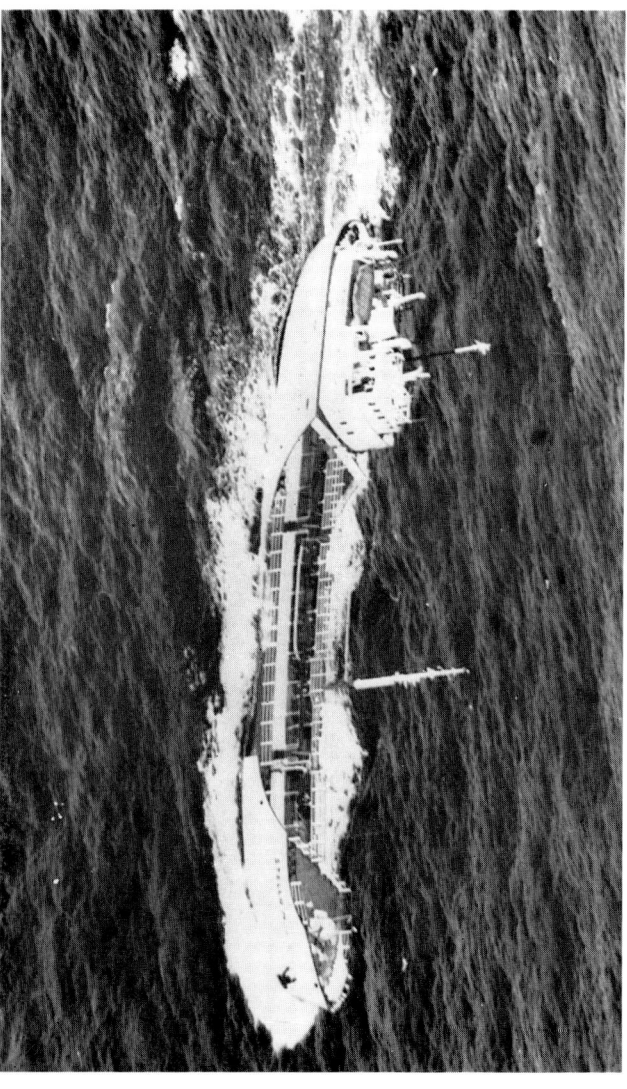

The Dutch ship *Stella Maris* on dumping voyage.

3. Degradation of the environment

POLLUTANTS found in industrial wastes vary greatly because of the diversity of industries and production processes involved. Refinery wastes include cyanides, heavy metals and chlorinated hydrocarbons while the wastes of the chemical industry contain arsenical and mercuric compounds. Contaminants in sewage, the principal domestic waste, are less diverse. They include organic materials, nutrient salts, silt and other suspended solids. Industrial wastes are usually more toxic than domestic wastes but in practice the two are frequently discharged together as industry has the right, under certain conditions, to pipe effluent into public drains.

Pollutants themselves vary in their chemical composition and behaviour, and consequently in the extent and the nature of their impacts on the environment. Some maintain their chemical composition for decades and even centuries whilst others are degraded into harmless materials in a matter of days. Some pollutants sink to the bottom. Others float or are suspended, and can be transported over huge distances. Horizontal and vertical water movements control their dispersion and dilution. Thus both the place and method of discharge is important in determining their behaviour and fate.

The greatest long-term danger from marine pollution lies in its ability to upset the ecological balance in the oceans. All marine life is joined in a complex web of food chains and the relationship between living things, and between living things and their surroundings, is so delicately balanced in the sea that a disturbance in one part can have a deleterious effect on the whole system. The long-term, low level effects of persistent wastes are complex and not well understood but the evidence is increasing that they are far more harmful than the more obvious short term impacts.

At the present time the only form of detriment to the marine environment which seems to be generally acknowledged is that which is self-evident and easily recognised – the fishless estuary, the oiled seabird, the muck and scum deposited on our beaches. But these are merely the outward and visible signs of more insidious and lasting damage.

Marine life

Some pollutants simply poison animals and plants with which they come into contact. Others make such a demand on the oxygen dissolved in seawater that higher life forms are eliminated. Some pollutants encourage the growth of single species which devour or poison other species. And others are concentrated by marine food chains to reach levels which upset the physiology and behaviour of predators at the head of the chain.

Lethal effects

Copper, zinc, cyanides, mercury compounds and pesticides are acute toxins. DDT, for example, can kill some forms of marine life in concentrations as low as one part DDT to 100,000m. parts of salt water. Mercury is actually refined by living things in the sea into a more deadly form – methyl mercury. Crude oil consists of a complex mixture of hydrocarbons and toxicity varies according to the mix and the source.

The poisonous fractions in crude oil are short-lived but oil pollution can also asphyxiate marine life mechanically. Heavy slicks will decimate seabird colonies, and clog or smother invertebrates in the tidal zone. The eggs and larvae of fish inhabiting surface layers of the sea are particularly susceptible to damage, and the gills of adult fish feeding near the top may also become clogged.

Detergents and dispersants used to treat spills emulsify the oil. It then ceases to be a mere surface layer and becomes an integral part of the aqueous environment where it can be ingested by zooplankton and other filter feeding organisms – with unknown consequences for the food chain to which they are linked. Experiments show that oil can also have a disastrous effect on phytoplankton. These microscopic plants are the basis of all primary production in the oceans. They are, so to speak, the grass of the sea on which all other life is ultimately dependent.

Bio-accumulation

Natural elements are normal constituents of unpolluted seawater and their concentrations have been reasonably constant for millions of years. Marine organisms at the base of the food chain extract these minerals from seawater and store them in their bodies. Their ability to concentrate materials within themselves varies from a few hundreds to several hundred thousand times the values in the surrounding sea solution.

Pesticides, synthetic detergents and organic chemicals are extremely stable substances and last for long periods in the sea without decomposing into harmless forms. These conservative materials can be concentrated by marine life in the same way as natural elements. This is also true of radioactive materials and of poisonous metals like lead, copper and mercury, whose natural occurrence in seawater is minute when compared to the concentrations at which they are discharged to the sea as wastes.

Organisms feed on plankton and successively pass on to higher organisms any substances absorbed. As this process moves through the food chain the contamination is amplified until toxic levels may be reached in predators at the head of the chain – including seabirds, sea mammals and man. The bio-accumulation of pollutants also has sublethal effects. Fish exposed to small doses of pesticides and copper show lower growth rates and reproductive activity. They are also more susceptible to disease.

Oxygen depletion

Oxygen in seawater supports marine life and is necessary for the biological degradation of organic materials since they provide an energy source for micro-organisms which use oxygen to respire. The dissolved oxygen in approximately 320,000 gallons of seawater is required to oxidise one gallon of crude oil completely. Thus the effect of dumping large amounts of organic matter in the sea is to reduce oxygen levels in the receiving waters.

In the open sea the powers of dilution are so massive that the dumping of sewage does not constitute a long-term problem but in confined, shallow areas marine life will suffer. Under conditions of gross pollution higher life forms disappear, and if the oxidation capacity of the dumping site is completely exceeded the area becomes anaerobic and devoid of all life above the level of bacteria. Oxygen deficit in a waste disposal area can become self-perpetuating as the accumulation of organic matter may act as a reservoir for future oxygen demand.

Over-enrichment

Sewage treatment aims to reduce de-oxygenation by breaking down matter before discharge but methods presently in use do little to remove the nitrates and phosphates in the sludge. Although these nutrients are harmless in themselves they can have deleterious effects on marine life because of changes they induce in the environment. An acceleration in the fertilization of plant life is caused which gives rise to phenomena called 'plankton blooms'.

Blooms consist of masses of planktonic organisms. They can spread over hundreds of square miles, liberating toxins which kill or stun fish, and poison shellfish. Even when the bloom is caused by non-toxic species of plankton their decay in shallow waters may reduce the diversity of marine life because of the large amounts of oxygen demanded in their decomposition.

Superfertilization occurs spontaneously in the marine environment. As a natural phenomenon it is quite common in tropical regions. In British coastal waters outbreaks appear to be on the increase. In the summer of 1968, a spectacular bloom was recorded off the north-east coast of England. It spread out from the mouth of the Forth and stretched all the way down the coast to Flamborough Head, Yorkshire.

Habitat changes

Nutrients, by changing the consistency of bottom sediments, can harm marine organisms living on the seafloor. Particulate trade wastes, dredge and mining spoils, by altering the bottom from sand or muddy ooze to hard mud, have the same effect. Individual species of sea animals are adapted to different types of habitat. When pollution alters the basic ecology of an area existing communities are driven out, and important spawning grounds may then be lost.

Materials consisting of finely divided particles also destroy habitats in shallow areas by masking the bottom from life-nourishing sunlight. Dumping close inshore will have a serious impact as suspended solids, instead of moving out and mixing with the open sea, are retained and recycled in coastal waters. Along the Durham coast, where colliery waste has been tipped, only 30 per cent of the species recorded there one hundred years ago now survive.

Large quantities of waste heat will also cause changes to occur in the distribution and abundance of marine populations as aquatic plants and animals have specific temperature requirements at different times of the year. Foreign species brought in on the hulls of ships become established, taking the place of less heat-tolerant native species. In extreme cases, where the temperature of the effluent exceeds survival limits, thermal pollution will eliminate flora and fauna. It can also increase the toxicity of certain pollutants already present in seawater.

Total disruption

Marine pollution imperils all forms of marine life by upsetting the ecological balance in the sea. Experiments show that small quantities of copper and zinc can prevent migratory fish from reaching their spawning grounds as they will not swim through waters which contain these pollutants. Oil also damages marine ecosystems. Many fish rely on the reception of chemical clues to select their habitats, food, mating partners, and to detect the presence of their enemies. Minute quantities of oil can apparently 'jam' these signals and trigger major biological disruptions.

In the laboratory it has been shown that man-made substances such as DDT, and poisonous metals like lead and mercury, impair the efficiency of photosynthesis. This is the process by which phytoplankton in the upper surface layers of the sea form organic compounds from water and carbon dioxide in the presence of sunlight. Photosynthesis fuels the entire life cycle in the oceans and any widespread interference with the process would have devastating results.

Human interests

The oceans provide a variety of goods and services, and pollution threatens their development and potential as a produc:ive resource. Floating refuse and surface films reduce recreational opportunities and damage aesthetic values. Public health problems are created by toxic agents and pathogens that find their way into seafoods. Economic loss is incurred when living resources are depleted or when fish are tainted and rendered inedible.

Marine recreation

The majority of all holidays in Britain are taken at the seaside and a large proportion of the population, including a high percentage of the retired, live permanently close to the sea. The coast is seldom more than a few hours drive away and with their growing mobility more people are attracted to the idea of spending a week-end by the sea. We use our six thousand miles of seaboard as an arena for swimming, sunning, surfing, water ski-ing, sport fishing, diving, boating, sailing. Millions enjoy these pleasures and pollution spoils their full enjoyment and use. Swimmers and scuba divers can no more exist in fouled, muddy waters than sardines or herring.

Many of our estuaries smell like oil refineries and our bathing beaches are continually coated by tar. Solid garbage and sewage refuse also create unsightly conditions and reduce amenity values. The Coastal Anti-Pollution League lists nearly two hundred resorts in England and Wales about which complaints have been received from the public and many local authorities acknowledge beaches in their areas to be unwholesome. Rotting algae and anaerobic waters, besides constituting a visual affront, produce pungent odours.

Depositional damage caused by offshore mining and dredging also leads to a general deterioration of swimming areas.

Recreation is only one amenity of the natural shoreline. Tidelands and wild-fowl populations in the coastal region have scientific and scenic value. Several nature reserves have been set up along the coast and some incorporate the seashore. No littoral habitats of ecological interest have been established yet but the Nature Conservancy is examining areas below low water mark, both inside and outside territorial waters, where conservation might be desirable on scientific grounds. A group of marine biologists in Devon have put forward a proposal for the conservation of an area of the seabed near Lundy Island. If accepted this will become Britain's first 'marine park or underwater reserve'. Obviously heavy pollution would jeopardise the success of such an enterprise as the assumption underlying wild-life conservation is that a particular complex of plants and animals can be managed in a way that leaves the basic character of the system unaltered.

Public health

Sewage contains pathogens and viruses, and outfalls located near bathing beaches are a potential health hazard. A study of sewage contamination in British coastal waters by the Public Health Laboratory in 1959 concluded that the risks to public health for all practicable purposes could be ignored – except in areas which were so obviously and grossly polluted that no one would think of swimming there anyway. These findings have not gone unchallenged and a more up-to-date survey of the problem at resorts on the Continent suggests that bathers in the sea run a high risk of contracting any one of nine diseases known to be caused by faecal pollution – including conjunctivitis, enteritis and typhoid.

Diseases can also be picked up by eating shellfish which have filtered pathogenic bacteria and viruses out of sewage contaminated waters. Hepatitis has been caused by oysters harvested from polluted areas and shellfish have been collected containing polio virus. Plankton blooms have been known to cause paralytic shellfish poisoning. Eighty-five people were taken ill after the heavy bloom along the northeast coast in 1968 had contaminated shellfish beds. Free swimming fish can also carry diseases picked up in anaerobic waters. Herrings, sprats, canned and smoked salmon have been identified as vehicles of infection.

The accumulation of toxic substances by marine food chains is now acknowledged as a major health problem. In the United States contamination by pesticides has rendered nine species of fish unfit for human consumption, and last year nearly a million cans of tuna fish were withdrawn from the market after it had been found that samples contained unacceptably high levels of mercury. This particular episode may have been a false alert but the fact remains mercury is a highly active poison which in low doses can disrupt the central nervous system – leading to madness and even death.

In the 1950's an outbreak of mercury poisoning occurred among people living in Minamata Bay, Japan. Their principal diet was fish and shellfish caught in the Bay. Doctors were baffled by its cause but eventually the source was traced to the discharge from a nearby chemical factory. Altogether one hundred and sixteen people were stricken. Forty-three died and many others suffered permanent disability. A second outbreak occurred from the same cause in Niigata in 1965, affecting thirty people and killing five. There have been cases of mercury poisoning in Sweden and people there are officially advised not to eat fish more than once a week.

Seafoods are becoming increasingly contaminated by oil

spills and concern has been expressed about the potential cancer risk. Cancerous growths have been found in a variety of free swimming fish, such as Dover Sole, and it is thought that oil polluted waters may be the cause. Research data from the United States confirms that hydrocarbons of the kind that can cause cancer in humans are concentrated by oysters and mussels in contaminated waters. No one has yet shown that cancer in man has resulted directly from the consumption of carcinogens in seafood but public health officials in America do not discount the possibility.

Submarine waste dumps can be injurious to health. In 1945 the Allies ditched 20,000 tons of German chemical warfare material in the southern Baltic. Fishermen handling nets and fish contaminated with the gas have been badly injured on several occasions. Although the containers were originally dumped twenty-five miles offshore in three hundred feet of water, the tides and currents have shifted them into shallower areas within a few miles of holiday coasts. Underwater currents and internal waves can make the bottom of the sea as restless as the top. In the summer of 1970, the Royal Navy dumped some ferric chloride containers near the Nab Tower, Portsmouth. A few weeks later several hundred canisters were washed up on the beaches of the Isle of Wight. Poisonous containers and old wartime mines seem to roll up on our beaches with remarkable frequency.

Economic damage

The waters overlying our continental shelf are among the most productive in the world. They include the nursery grounds for a number of economically important fish. They also support shellfisheries for oysters, mussels, clams and shrimps in protected inshore waters and estuaries, and lobster fisheries along rocky coasts and inlets. Restrictions placed on shellfish cultivation and on the sale of fish are a heavy and continuing loss for the fishing industry.

The Public Health (Shellfish) Regulations 1934 allow local authorities to make orders that prohibit the sale of shellfish from grounds which have been polluted unless they are cleansed, sterilized, or relaid in clean water. The value of shellfish production destroyed or impaired by pollution cannot be known as the contamination has been spreading for years but under the regulations many grounds have been closed and production elsewhere has been severely restricted. Oyster fisheries in the Tamar and Lynham rivers have been stopped completely, and cultivation in the rivers Colne, Blackwater, Roach, Whitstable and in Poole Harbour, is badly affected. Orders have also been placed on mussel fisheries at Lytham, Exmouth, Morecambe Bay and the Wash.

Visible fish kills in the sea are comparatively rare. Mature fish are so debilitated before death that they fall easy prey to predators and scavengers. Evidence with which to measure the extent of any damage done is then lost. A number of commercial fisheries, including such economically important fish as plaice, spend the early part of their life as eggs or larvae in the top five centimetres of the ocean. At this stage they are very vulnerable but again evidence with which to measure the extent of any destruction would scarcely be visible.

The loss of shellfish beds, the obliteration of habitats by dredging operations, the destruction of salmon and trout fisheries caused by pollution barriers across bays and estuaries, should in theory be capable of some economic evaluation. Otherwise the long-term, low level effects of pollution on living resources of the sea are so complex, subtle and diverse that they defy measurement. Mortality, reproduction, growth and the movement of stocks can all be affected. Quality may also be impaired. Heavy metals discolour shellfish and diminish their value. Tainting will reduce their market value as well, always assuming that they can be sold in that state at all. Experiments have shown that the oily residue escaping from a single outboard motor is sufficient to spoil the flavour of all fish within one acre-foot of the engine.

The ambivalent nature of many pollutants makes any assessment of the damage doubly difficult. Some 'pollutants' have beneficial effects. The china clay residues in St Austell Bay, for example, have altered the flora and fauna of the seafloor. As a result lobster and crab fisheries have been lost. Yet other local fisheries show a remarkable improvement. In rich sea areas superfertilization may be destructive but in barren or unproductive regions a nutritious outfall, carefully sited, could be an asset. Likewise, waste heat from power stations can be employed profitably as the warm water accelerates the growth of young fish. The thermal effluent from the Hunterston Power Station is in fact exploited for a fish-farming experiment. Dumped junk may be dangerous but equally artificial reefs made up of old car hulks make surprisingly productive fisheries.

The fishing industry also loses out when pollution interferes with fishing operations. Dredging can leave the seabed in a state unsuitable for trawl and seine fishing. Sunken oil is a menace to bottom fishing as the tarry lumps are caught in the nets and foul the gear. Synthetic ropes, net twine and wire hawsers foul propellers and are a recognised navigational hazard. A new danger for small fishing vessels has appeared recently in the shape of polythene sheeting. These are virtually indestructable in the marine environment and have become a notable problem in the Irish Sea. If the material sinks or remains adrift in mid-water it is a potential hazard to trawling since it can prevent the flow of water through the meshes and wreck the gear. The chief danger however lies in their being sucked into cooling water intakes and overheating the engine at a critical moment in navigation.

No inventory of the economic consequences of marine pollution can be complete without an eye on the future. Contaminated seawater could seriously impair the efficiency of desalination plants and affect the economics of turning salt water into fresh. Fish-farming is obviously conditional on keeping designated areas pollution-free as successful hatcheries require very pure water conditions. Unexploded projectiles on the seafloor foreclose areas to dredging and mining operations. Mining minerals from the seabed is difficult enough without having to armour-plate the machinery against booby traps. The use of the seafloor as a junkyard proves equally exasperating to research vessels towing expensively designed equipment.

4. The scope of present laws

THE SOVEREIGNTY of a state extends beyond its land territory to a belt of sea adjacent to its coast described as the territorial sea. There is no internationally agreed limit to this area but Britain herself claims three miles. Territorial waters, however wide, are measured from baselines drawn around the coast. The normal baseline is low water mark but more detailed provisions are set out in the Geneva Convention on the Territorial Sea 1958. Waters landwards of the baseline form part of the internal waters of the state.

The Territorial Waters Order in Council 1964 established the baselines for Britain. Between Cape Wrath and the Mull of Kintyre a series of straight lines joining specified points lying to the seaward side of the Outer Hebrides were used, and closing lines not exceeding twenty-four miles in length were drawn across bays and estuaries like the Moray Firth and the Bristol Channel.

The Geneva Convention on the Continental Shelf 1958 also gave coastal states the exclusive right to explore and exploit seabed resources in submarine areas adjacent to their coast out to a depth of two hundred metres, or beyond that limit to where the depth of the superjacent waters admits of exploitation. Where the same shelf is adjacent to two or more states whose coasts are opposite one another the boundary is settled by agreement. The median line is usually chosen and this is calculated from the same baseline used to measure the territorial sea.

The marginal seas around Britain overlie an extensive area

Territorial sea and shelf boundary

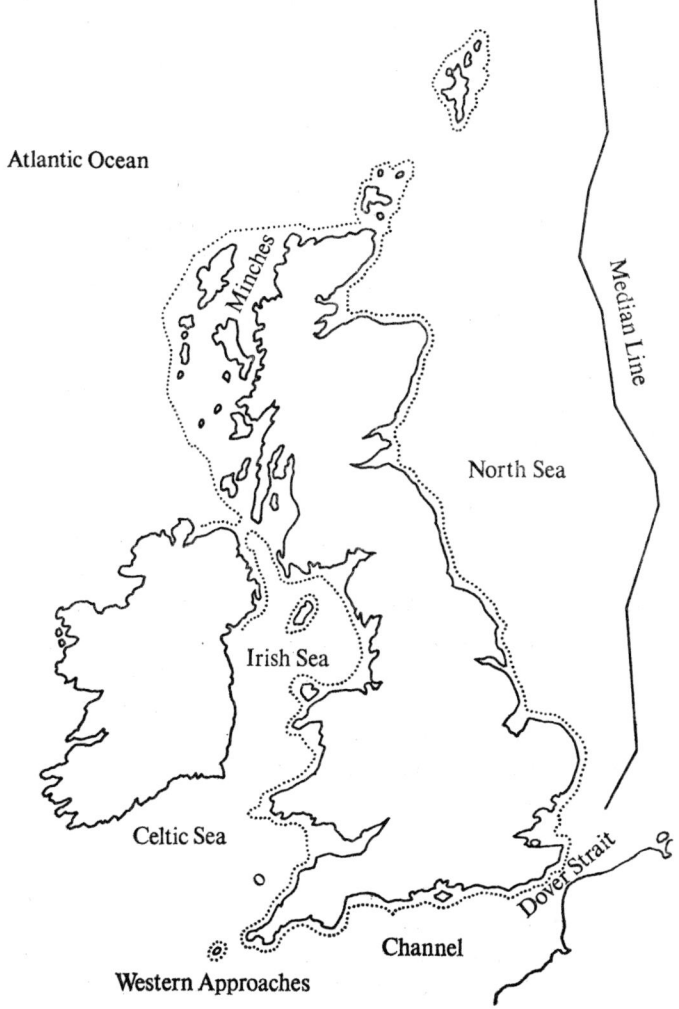

Atlantic Ocean

Minches

Median Line

North Sea

Irish Sea

Celtic Sea

Channel

Dover Strait

Western Approaches

of continental shelf, and the provisions of the Convention were brought into effect by the Continental Shelf Act of 1964 and a series of bilateral treaties. In the North Sea, which is all shelf, Britain shares a common boundary with Norway, Denmark, Germany, Holland, Belgium and France. In the Channel, the Irish Sea and the Atlantic Ocean no firm underwater frontier has yet been drawn. The area of seafloor subject to British jurisdiction, calculated to the two hundred metre isobath, is approximately 180,000 square miles.

Inside the territorial sea Britain has the right to control pollution from any source – including foreign vessels. On the continental shelf we can control all installations engaged in the exploration or exploitation of resources lying within our area of jurisdiction. The waters overlying the continental shelf are classed as 'high seas' and no country may validly claim sovereignty over them.

Voluntary pollution

In Britain there is no authority specifically charged with the conservation of the marine environment but some pollution abatement powers are vested in the River Authorities and Sea Fisheries Committees. River Authorities control rivers and estuaries, and the Sea Fisheries Committees waters along the open coast. Their jurisdiction does not reach beyond the territorial sea and there are no statutory controls over dumping outside these limits except in the case of radioactive wastes and oil.

Pipeline disposals

The Rivers (Prevention of Pollution) Act 1951, for maintaining or restoring the wholesomeness of rivers, applied pollution controls to non-tidal rivers but the Minister had power to make an Order on the application of a River Authority to

Colliery tipping on the Durham Coast.

The author preparing to dive in the submersible *Pisces*.

give control over new discharges to estuaries. In 1960 more general powers were granted under the Clean Rivers (Estuaries and Tidal Waters) Act to control new and substantially altered discharges up to the seaward limit of certain estuaries. The Secretary of State for the Environment (or for Wales) also has power under the Rivers (Prevention of Pollution) Acts 1951 and 1961, on application by a River Authority, to make 'Tidal Waters Orders' which give full control over both new and existing discharges in marine waters.

Under these Acts the River Authorities can impose conditions regarding the treatment, composition and quality of the effluent before granting consents to discharges, and they can inspect outfalls to ensure that conditions are observed. The Salmon and Freshwater Fisheries Act 1923 gives River Authorities some additional controls over marine waters. These are of limited value as they relate only to liquid or solid substances which are injurious or poisonous to salmonoid fish, their spawning grounds or food. No account is taken of detriment to the marine environment itself.

The Sea Fisheries Regulation Act 1966, which repealed and re-enacted a number of previous enactments, empowered Sea Fisheries Committees to make bye-laws for the purpose of prohibiting or regulating the deposit or discharge of any solid or liquid substance detrimental to sea fish or sea fishing. The interpretation of these powers varies among the twelve Committees in England and Wales. A number of them take what could be called an 'ecological viewpoint' in which 'detriment' is taken to include any environmental change that could adversely affect the viability of fish stocks. A new Byelaw which the Lancashire and Western Joint Sea Fisheries Committee have submitted to the Ministry of Agriculture, Fisheries and Food for confirmation seeks to codify this ecological approach.

The powers of the Sea Fisheries Committees are defective in a number of respects. They cannot be employed directly to control threats to public health or amenity. They are subordinate to the power of River Authorities to grant consents to discharges and to the statutory powers of local authorities to discharge sewage. In these cases the Committees only have the right to be consulted and this is rarely more than nominal.

This subordination reduces much of the effectiveness of the existing legislation as rivers and estuaries are by far the biggest source of pollution at sea. Sewage outfalls also carry large quantities of trade wastes and existing regulations create the anomaly that industrial effluent is subject to external controls whereas the same wastes carried in sewage are not. The Public Health Act 1936 requires local authorities to carry out their functions so as not to create a nuisance. They also have to obtain the sanction of central government for loans required for capital expenditure on sewage disposal schemes. In practice approval is not given unless the scheme avoids risks to health and amenity.

There are no Sea Fisheries Committees in Scotland although the law does make provision for them. Controls are vested in river purification authorities. Powers granted under the Rivers (Prevention of Pollution) (Scotland) Act 1951 are similar to those governing the control of new discharges in England and Wales. The Rivers (Prevention of Pollution) (Scotland) Act 1965 widened this to cover all discharges predating the 1951 Act and also gave jurisdiction over new discharges in certain 'controlled' tidal waters. These embrace all coastal areas where pollution is most likely to occur. Where it is shown that stricter control of tidal waters is required the Secretary of State for Scotland may, by Order, extend the full powers of both Acts to any of the tidal waters around the Scottish coastline.

Sea fisheries districts in England and Wales

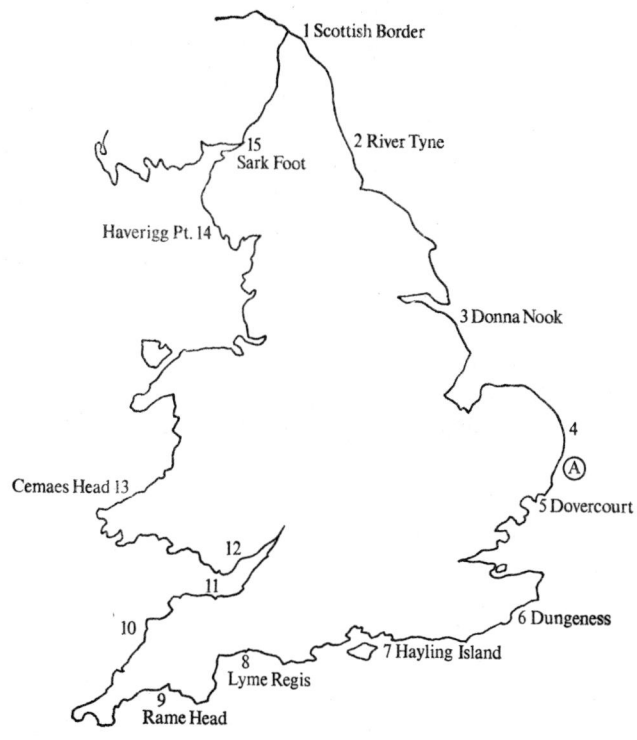

1 Scottish Border
2 River Tyne
15 Sark Foot
Haverigg Pt. 14
3 Donna Nook
4
(A)
5 Dovercourt
Cemaes Head 13
12
11
6 Dungeness
10
8
7 Hayling Island
Lyme Regis
9
Rame Head
♂ 16

Northumberland (1–2)	Cornwall (9–10)
North Eastern (2–3)	South Wales (12–13)
Eastern (3–4)	Lancashire & Western (13–14)
Kent & Essex (5–6)	Cumberland (14–15)
Sussex (6–7)	Scilly Isles (16)
Southern (7–8)	East Suffolk &
Devon (8–9 & 10–11)	Norfolk River Authority (A)

Radioactive waste is in a category of its own. The Radioactive Substances Act 1960 provides that no account is to be taken of radioactive materials for the purposes of the Acts mentioned above, and subjects disposals to ministerial authorization. The Department of the Environment and the Ministry of Agriculture share responsibility. Radioactive materials are the most carefully controlled of all wastes and coastal discharges are usually kept well below the minimum safety levels recommended by international organisations. The Ministry of Agriculture, Fisheries and Food keep a check on radioactive levels in seawater and ensure that authorizations are followed.

Shipborne dumping

Inside territorial waters the Sea Fisheries Committees have power to regulate the dumping of waste from ships where this is detrimental to sea fish or sea fishing but outside the three mile limit there is no authority whose consent is legally necessary. A voluntary system, however, is operated by MAFF (DAFS in Scotland) under which waste disposal companies agree to keep to designated spoil grounds and to cooperate in effluent tests. The Ministry recommend the area and the method of discharge. The toxicity and persistence of the materials is considered as is the dilution and dispersion available at the dumping site. In the case of highly toxic wastes companies are advised to dump in sealed containers on the deep seafloor beyond the continental shelf.

Each dumping is entered in the ship's log and a certificate containing details of the materials and of the dumping site is signed by the master and filed. But there are no sanctions and only a casual testing system is applied. Samples are taken at the dockside by fishery officers and sent to the Marine Pollution Unit at Burnham-on-Crouch. This is done about three times a year. More regular checks are made by the dumping companies themselves. Purle Brothers, for example, try to test 10 per cent of their cargoes. The analyses however take upwards of fourteen days and by this time the load in question has frequently been dumped.

Britain is a signatory of the North Atlantic Fisheries Convention 1967 which states, among other things, that no vessel shall throw overboard anything which can hinder fishing, or harm fish, fishing equipment or fishing vessels. But 'vessel' is defined as meaning only fishing vessels. Britain has also signed the Geneva Convention on the High Seas. Article 25 lays down that every state must take measures to prevent pollution of the seas from dumping of radioactive waste, taking into account any standards or regulations which may be formulated by competent international organizations.

There are in fact no international regulations *per se*, but there is what amounts to an international code of practice. Recommendations made by the International Commission on Radiological Protection are observed by most countries, and the International Atomic Energy Agency has recently suggested that a manual of dumping methods should be written and an international register of disposals compiled.

The Radioactive Substances Act 1960, by making it an offence to use radioactive materials without registration, creates what amounts to a licensing system for the dumping of atomic wastes at sea. Methods of packaging, handling and transport are subject to approval and inspection. Dumping is carried out in approved areas which are in water depths of at least one thousand five hundred fathoms. Containers are designed to reach the seabed intact, each dumping is supervised, and an extract from the ship's log is filed.

The Navy make their own dumping arrangements for unwanted stores of a dangerous nature like corrosive chemicals and explosives. The dumping of surplus stores has recently been stopped. The dumping of ammunition and explosives is still allowed and the policy is to confine disposals to as few places as possible and in as deep water as possible. Consideration is given to the dumping grounds being clear of fishing grounds, submarine cables and anchorages. They are all marked on Admiralty charts.

Operational discharges

Since 1922 British law has prohibited the discharge of oil or oily mixtures into our territorial waters from ships and shore-based installations. Operational discharges on the high seas are subject to the International Convention for the Prevention of Pollution of the Sea by Oil 1954, as amended in 1962 and 1969. The Oil in Navigable Waters Acts 1955, 1963 and 1971, as consolidated in the Prevention of Oil Pollution Act 1971, gives effect to the Convention so far as British ships are concerned.

The 1954 Convention prohibited tankers from discharging persistent oil or a mixture containing one hundred parts per million or more of such oil within certain prohibited zones. Outside of these zones, however, there were no restrictions on tankers, and in the case of ships other than tankers the only requirement was that they should discharge as far as practicable from the land. Under the amendments of 1962 the prohibited zones were extended to one hundred miles along many coasts, and the whole of the North Sea together with a large part of the eastern Atlantic became a prohibited area. Also ships above twenty thousand tons built after the coming into force of the amendments were prohibited from discharging anywhere at sea, and not only in prohibited zones, if the oil content of the effluent exceeded one hundred parts per million.

Further amendments agreed in 1969 will replace the prohibited zone system by a permitted rate of discharge. Henceforth tankers may discharge oil at sea provided that the instantaneous rate of discharge does not exceed sixty litres per mile of distance travelled, and on condition that the total amount of oil dumped does not exceed 1/15,000th of the total cargo carrying capacity during any one voyage. Tankers however, must be more than fifty miles from land when discharging. For ships other than tankers the same permitted rate of discharge is allowed except that the oil content in this case has to be less than one hundred parts per million. The ship must also be as far as practicable from the land.

The enforcement of restrictions is aided by the keeping of record books in which the ship's master must account for the disposal of oil cargoes and all oily residues. Surveyors from the Department of Trade and Industry carry out inspections of ships to make sure that they comply with requirements, and the Oil in Navigable Waters Act 1971 has substantially increased the fine for violations. Illegal discharges from underwater exploration and exploitation activities on the continental shelf are governed by the Continental Shelf Act 1964. The new Act of 1971 further strengthens these provisions, and fills a gap in the law by extending the prohibitions on operational discharges to activities on the seafloor inside the territorial sea.

Responsible shipping organizations and oil companies have made considerable efforts voluntarily to control oil discharges from ships. The 'load-on-top' system, pioneered by Shell and first introduced by them in 1962, allows the oily residue remaining after the discharge of ballast and cleaning water to be retained on board for offloading at ports. About eighty per cent of the world's tanker fleet now use this system and it is estimated that over a million and a half tons of oil per annum, which would previously have been ditched at sea, are now saved.

Involuntary pollution

Ships are subject to the nationality of the flag state and Britain has full control over her own ships anywhere at sea. She has no jurisdiction over ships flying a foreign flag on the high seas except where international treaties provide otherwise. But foreign vessels exercising the right of innocent passage in our territorial waters must comply with our laws and regulations, especially those concerned with transport and navigation. Passage is not considered innocent if it is prejudicial to the peace, good order or security of the coastal state.

Shipping casualties

The Department of Trade and Industry has responsibility for all matters concerned with shipping and navigation. Freedom of navigation on the high seas, however, makes safety at sea an international problem and collaboration is promoted by the Inter-Governmental Maritime Consultative Organization. Before, but particularly since, the wreck of the Torrey Canyon, IMCO has been busy discussing ways and means of suppressing accidents in crowded seaways. Their review covers: revision of the International Regulations for Preventing Collisions at Sea, lights and markings to facilitate passage, obligatory navigational aids, pilotage, routeing in narrow sea lanes, ship maneouvrability and design, training and seamanship standards, hydrographic surveys, wreck dispersal, and offshore tidal gauges to provide 'real time' information on tide heights.

In 1964 IMCO agreed on a traffic separation scheme for the Dover Strait. This came into force in 1967. The scheme until now has been voluntary and a recent census of channel shipping reveals that at least ten per cent of the ships using the Strait ignore the recommended routeing. This is equivalent to forty ships a day. Early in 1971 a Norwegian chemical carrier hit the Nab lighthouse six miles off the Isle of Wight and later a German coaster collided with the Mid-Barrow Lightship moored ten miles off Clacton-on-Sea. The number of near misses in the well lit and well advertised Texaco Caribbean/Brandenburg graveyard shows how difficult it is to secure compliance with international regulations outside territorial waters.

In Britain the Standing Advisory Committee on the Carriage of Dangerous Goods in Ships advises the Department of Trade and Industry which substances should come within the scope of the Merchant Shipping (Dangerous Goods) Rules 1965 and makes recommendations as to how such goods should be packaged and stowed so as to minimise the risks. The advice is published in the 'Blue Book'. This has no statutory force. Moreover, the rules have been designed to protect the crew and passengers. With the exception of radioactive substances, little account is taken of the risks of spillage following a collision or stranding. Nor are there any special rules governing the navigation of ships carrying noxious cargoes.

The Department of Trade and Industry, with the co-operation of the armed forces, is responsible for combatting oil pollution at sea. Their organization is based on the Marine Survey Service whose principal officers decide what action should be taken. Local authorities, with the advice and assistance of the central government, are responsible for cleaning up beaches and for spillage up to about a mile offshore. The Coastguard Service provides an early warning system and all British ships and aircraft are asked to report any incident likely to result in pollution of the coasts. There is also a regional agreement in force under which North Sea states have agreed to cooperate in reporting and tracking major oil slicks.

The International Convention on Civil Liability for Oil Pollution Damage makes tanker owners liable for oil pollution damage and advances the position of claimants. Victims of pollution can claim compensation up to higher limits and in wider circumstances than at present, covering both clean up and preventative costs. The Convention was signed in 1969 and ratified by Parliament in the Merchant Shipping (Oil Pollution) Act 1971. Tanker owners are held to be strictly liable, with exceptions, and certificates of insurance have been made compulsory.

Some additional compensation has been provided by industry through the Tanker Owners' Voluntary Agreement Concerning Liability for Oil Pollution (TOVALOP) and the Contract Regarding an Interim Supplement to Tanker Liability (CRISTAL). These voluntary agreements cover physical damage to the land adjoining waters navigated by the tanker concerned. They do not cover fire, or explosions, consequential damage or ecological impairment. Nor do they cover cargoes other than oil. IMCO is examining the possibility of setting up an international fund to meet any part of the clean up costs not covered by the 1969 Convention. This would replace the voluntary arrangements.

Offshore installations

The soil of the sea as far as territorial limits is vested in the Crown. This includes the bed of all estuaries and arms of the sea, and of all navigable rivers up to the point where the tide flows and reflows. The Continental Shelf Act also vests in the Crown all rights exercisable in the UK share of the seabed beyond territorial waters. The working and getting of coal is an exception. Exclusive rights both inside and outside territorial waters have been granted to the National Coal Board.

The Department of Trade and Industry controls and supervises the conditions under which oil and natural gas are exploited in offshore areas. The Department's consent must also be obtained before the NCB can engage in any mining operations on the shelf. This may be given on such terms and subject to such conditions as the Minister thinks fit. The exploitation of mineral resources other than coal, oil and gas is supervised by the Crown Estate Commissioners.

Before granting a licence for the winning of sand and gravel the Commissioners seek the advice of the Department of the Environment about the effects of dredging on coastal erosion, and of the Ministry of Agriculture about the danger to fisheries. In some areas around Britain dredging is not allowed because of the risk of damaging, directly or indirectly, the feeding or breeding habitats of commercial fisheries. Under the Coast Protection Act 1949 the Minister of Trade and Industry has powers to prohibit seafloor mining in coastal waters if the operations are likely to obstruct or hinder shipping.

Article 5 of the Convention on the Continental Shelf lays down certain regulations for the safety of activities on the seafloor. Subject to the overriding requirement that exploitation must not unjustifiably interfere with recognised shipping lanes, coastal states may build and maintain any structures on the continental shelf and establish safety zones around them out to a distance of five hundred metres. Due notice must be given of the construction of any installations, and permanent means for giving warning of their presence must be maintained. Inside these safety zones parties to the Convention must take all appropriate measures for the protection of living resources of the sea from harmful agents.

The Continental Shelf Act 1964 and the Petroleum (Production) (Continental Shelf and Territorial Sea) Regulations 1964 give effect to these provisions. Mining operations require the consent of the Department of Trade and Industry if any obstruction or danger to navigation is involved (s.4) and for the purposes of protecting installations the Minister may prohibit British ships from entering any part of a designated area. Masters or owners of ships who fail to comply are liable to a fine or term of imprisonment (s.2). Acts or omissions on, under, or above offshore installations or inside safety zones are subject to UK laws (s.3).

Section 16 of the Regulations stipulates that operators must prevent the escape or waste of petroleum, and avoid damage to any oil bearing strata. The Regulations also contain model clauses for inclusion in individual licensing agreements. Oil rigs are provided with at least two 'blow-out preventors' and other 'fail safe' techniques are employed. Department of Trade and Industry petroleum inspectors ensure that regulations are followed.

The Geneva Convention on the High Seas requires signatories to draw up regulations to prevent pollution of the sea by the discharge of oil from pipelines, and to introduce legislation making the breaking or injury to pipelines a punishable offence. These provisions were put into effect by the Continental Shelf Act 1964 which extended the Submarine Telegraph Act 1885 to include pipelines under the high seas. All pipelines and rigs are marked on Admiralty charts warning mariners not to anchor or trawl in the vicinity. The Hydrographer of the Navy also issues warnings in Notices to Mariners published annually. Department of Trade and Industry inspectors receive reports of pipeline inspections from operators who carry out regular surveys.

5. Towards a clean ocean

THE RIGHT to dispose of waste materials in the oceans is one of the traditional freedoms of the high seas but this freedom, like all other freedoms, must be exercised with reasonable regard to the interests other countries have in the marine environment. Any activity which unjustifiably interferes with another current or potential use of the oceans could be unlawful. Unfortunately, the law of the sea does not establish a hierarchy of ocean uses which can be used to settle conflicts in resource use, and whether or not a disposal scheme conforms to the standards of international law will depend on the particular circumstances of each case.

Discussions on marine pollution tend to place too much emphasis on the capital costs of treatment plant or alternative means of disposal, and too little emphasis on the true value of other uses and resources which may be damaged irrevocably. In an island as small and crowded as Britain it would be unreasonable to insist that our coastal seas should remain completely unpolluted but it is not unreasonable to expect that sources of pollution will be controlled so as to protect the interests of other users and avoid needless damage.

The management of waste disposal

The Working Party on Sewage Disposal, commonly referred to as the Jeger Committee, recognised that our present laws were inadequate and called for a comprehensive system of controls. In their report the Committee recommended that these controls should be exercised by re-titled River Authorities. The obvious solution would have been to enlarge the powers of the existing Sea Fisheries Committees but their claim was rejected on the grounds that the River Authorities had more extensive technical facilities, as well as long and wide experience of pollution problems.

This is a strong point as only one Sea Fisheries Committee (the Lancashire and Western) is equipped with its own pollution staff and research laboratory. From a scientific point of view, however, an equally good case can be made for having separate authorities to deal with the problems of pollution within the maritime belt. Limnology, from whose ranks the scientists of River Authorities are drawn, is a totally different specialisation to oceanography. In the oceans conditions and processes are infinitely more complex than in freshwater environments, and their study requires a different approach and different techniques.

Coastal sea authorities

The seaboard of this country is put to many uses other than waste disposal – sport and commercial fishing, recreation of all kinds, navigation and sea transport, wild-life conservation, mining, water storage and abstraction, to name just a few – and what is really wanted is some organization with authority to preserve, and where possible to enhance, the productive capacity of this region as a total resource. This would entail granting powers to control marine disposal, to vet all developments and activities in the coastal zone, and to adjudicate between different uses and users when these are in conflict.

Local authorities, rather than central government, are usually given the power to control waste disposal as they are made more quickly aware of nuisances and hazards. They are also more directly conscious of any local facilities available.

The problems of sea pollution, on the other hand, are more regional in character as they are conditioned by configurations of the coastline and water movements centred round them. The effectiveness of any new administrative units would be diminished if naturally occurring divisions along the coast were to be broken up artificially. The financial and research needs of any new statutory bodies also need to be taken into account when considering lower limits to size.

The boundaries of River Authorities have been drawn to coincide with major rivers and their tributaries, and the coastline within the jurisdiction of individual authorities bears no relation to hydrographic conditions existing offshore. The seaboards of the twelve Sea Fisheries Committees, which cover most of the coastline of England and Wales, are altogether more suited to the purposes of pollution abatement and it is suggested that, properly reconstituted, they would form a much better basis for any new system of controls.

The Sea Fisheries Committees are committees of local authorities, not representative bodies. They are financed from precepts levied on councils within the districts, and their byelaws require the confirmation of the Minister of Agriculture. The Committees are made up of members of the constituent local authorities, representatives from River Authorities within the district, and people appointed by the Minister as being acquainted with the needs and opinions of the local fishing industry.

These Committees should be made directly responsible to the Secretary of State for the Environment, and renamed Coastal Sea Authorities. Their powers should be extended to cover all discharges in the territorial sea, including estuaries. When formulating a policy of environmental control the Authorities must be free to interpret their powers as covering all present and future uses of the maritime belt. Obviously adequate time would have to be given for required standards to be reached in the case of existing discharges.

In addition to these direct controls the new Authorities should act as environmental assessors. Before any licensing or planning authority grants permission for industrial development along the coast, or on the seafloor itself, they should be required by law to submit the scheme to the local Coastal Sea Authority for an environmental impact report. These reports should be regarded as an integral part of all planning procedures and appeals.

Membership could be made up in the same way as the existing Fisheries Committees except that all users of the coastal zone must be included and appointments would be made by the Department of the Environment. New funds could be provided by charging a fee for disposals. This should be levied at a rate sufficient to cover research and monitoring costs. If ships using British ports are required to pay port dues, and offshore gravel operators royalties, there can be little justice in allowing the sea to be exploited as a 'free' resource for waste disposal purposes. An economic charge would make it easier to judge the relative merits of land as against sea disposal. Companies might also be encouraged to investigate the recycling and reuse of many wastes.

At present the onus is on the controlling authority to prove that the discharge under consideration will cause harm to the environment. It would be a great improvement if the burden of proof was to be reversed so that the discharger had to show that there would be no detriment to the environment, or justify why his use for the area in question should take precedence over other uses. Too often if the water is already polluted it is taken by the would-be polluter as a licence to discharge *ad lib*. Where the burden of proof rests with the discharger he is encouraged to follow a more enlightened course.

Licences for dumping

The Jeger Committee also recommended that the Government should take statutory powers to control dumping by British ships or ships using British ports. The Technical Committee on the Disposal of Toxic Solid Wastes, on the other hand, considered that the voluntary system for controlling dumping beyond the territorial sea had worked well, and were opposed to any change.

The voluntary system has worked well and this is due in no small measure to the integrity of the companies involved. Nevertheless, ocean dumping under present procedures is not subject to fool-proof tests for toxicity and the practice is bound to grow as anti-pollution laws on land are strengthened. Without statutory powers the government has no legal means of bringing recalcitrant companies to book, and victims of pollution damage could find themselves without redress. Several companies have declared themselves in favour of controls.

Studies made of the dumping grounds in the outer Thames Estuary and the Firth of Clyde reveal marine populations in the areas as almost normal. But these studies aimed to show the gross local effects only. The pattern of distribution of the materials dumped is not known, nor are their low-level, long-term effects. It must be a matter of concern that among the millions of tons of sewage sludge dumped each year there are several hundred tons of heavy metals and, according to the Freshwater Fisheries Laboratory at Pitlochry, at least one ton of PCB's in each area.

Controls at home would also conform to practice abroad. Sweden, Denmark, Norway and Iceland have decided jointly to ban the dumping of certain toxic products by their firms in international waters as from January 1972. Finland and Holland already exercise controls. And in the United States there is to be legislation to ban unregulated dumping of all materials, with strict procedures of enforcement and control.

It also seems that we are under an international obligation to control dumping practices. Article 25 of the Geneva Convention on the High Seas states 'all states shall cooperate with the competent international organizations in taking measures for the prevention of pollution of the seas . . . resulting from any activities with radioactive materials and *other harmful agents*'. This clause has been invoked to support the case for international action and an agreement has been drafted to regulate marine disposals in the north-east Atlantic area which sets out certain principles and criteria to govern ocean dumping.

It is not necessary for Britain to sit on her hands waiting for international developments. Coastal states have full control over their own and foreign vessels inside the territorial sea, and they can control the transport of wastes from their own ports. A compulsory system should be adopted which will allow the Government to ban the dumping of specific materials and to designate safe sites.

All types of dumping should be controlled, including that done by the armed services. The same criteria should be used in assessing degradation to the environment as is to be used by the new Coastal Sea Authorities, and again a fee for disposals ought to be charged. Biologically productive areas must be delimited and protected. Responsibility for operating the system would fall most appropriately to the Department of the Environment but provision should be made for consultation with interested parties including the Ministry of Agriculture, the Hydrographer of the Navy, Department of

Trade and Industry and the Institute of Geological Sciences.

Controls can be exercised at the loading point rather than at the dumping grounds, and the day-to-day administration of the system could be carried out by the Coastal Sea Authorities from ports within their districts. Officers of the CSA's might escort the ships to the spoil grounds from time to time, and conduct spot checks. At other times a watch could be maintained by coastguards, lightships, fishery protection vessels, and other units used in the general policing of the marine environment. With record books and adequate sanctions there would be no real difficulty in enforcing restrictions.

The Government already has power under the Radioactive Substances Act to provide facilities for the disposal of radioactive wastes and consideration should be given to extending this to other difficult or dangerous products. There has been a suggestion that toxic wastes could be disposed of permanently by dumping them in deep sea trenches. Convection processes in the earth lead to the upwelling of materials from the earth's crust in some parts of the seafloor. This is balanced by a similar descent of materials into the earth's interior. Wastes deposited in the neighbourhood of these 'tectonic sinks' would be withdrawn from the seabed.

Banning oil discharges

The Sea Fisheries Committees have power to prosecute in the case of offences against the Oil in Navigable Waters Acts. No purpose would be served in transferring the responsibility for checking the disposal of oily residues and tank washings from the Marine Survey Service of the Department of Trade and Industry as it has other duties concerning seaworthiness, safe operations and the manning of ships. But responsibility for beach cleaning and oil spillage in the coastal zone would fall naturally to the new Coastal Sea Authorities.

The 1954 Conference on Oil Pollution passed a number of resolutions of which one of the most important declared that a complete ban on the discharge of persistent oils should be observed from the earliest practicable date. This should remain the goal and in view of past procrastination by the shipping industry the Government should now set a firm date. The new permitted rate of discharge may or may not be innocuous in the context of beach pollution but it will not prevent damage to the marine environment itself. Nor are the amounts of oil involved negligible. At these rates at least 100,000 tons of oil will continue to reach the sea from tankers alone. There will be quite enough spillage from shipping casualties without having to contend with wilful discharges.

The prevention of accidental pollution

The threat of oil pollution has been a major consideration in international discussions on improving safety at sea. Hazardous cargoes other than oil, however, have not received the attention they deserve. The danger was highlighted by the Rhine fish kill in which about two hundred pounds of the pesticide endosulfan killed at least one hundred tons of fish. There was a less well known incident recently off the Spanish coast. A ship carrying 1800 drums of dieldrin mixed with mercury compounds ran aground near Corunna. Five hundred drums were lost overboard. A £2m. local oyster industry was wiped out and marine life in the area devastated. Fortunately, strong currents near the coast prevented a wholesale catastrophe.

Oil rigs and other offshore installations have also been neglected. Although they come under the jurisdiction of coastal states, they do not possess the status of islands and have no territorial sea of their own. Consequently states can only exercise such powers in 500 metre safety zones as international laws allow. These are not adequate. Ships have been known to pass straight through oil complexes in the North Sea, and a BP rig was actually rammed on one occasion. The situation has added urgency now that oil and gas concessions are to be granted in the Western Approaches, right in the path of channel shipping.

Advances in engineering may eventually diminish the danger of accidents. Several oil companies are developing subsea production systems which will obviate the need for platforms on the surface. Work is also in hand on new techniques which will allow the permanent burial of pipelines at depth. In the meantime IMCO should review the adequacy of the 500 metre safety zone, and consider what further steps can be taken to prevent trawlers habitually fishing inside complexes, and up and down pipelines exposed on the seabed.

Sea traffic control

IMCO has recommended that member states should make the traffic separation schemes in the Channel mandatory. Other measures are to be taken shortly but it remains to be seen whether the new rules and procedures can be satisfactorily enforced on an international basis. A large number of maritime states are involved, and many have no incentive to comply. Admittedly, ship owners and masters have a primary interest in safe navigation but they are also under strong pressure to meet financial deadlines. There is no doubt that many will take the quickest way for the quickest return – racing tides, cutting corners, crossing sea lanes and shelving maintenance schedules.

In their Charter for Channel Safety the Labour Party has proposed that Britain should extend her territorial limits to the median line in the Channel so that she can apply her own regulations and standards in these waters. While the Government has not completely ruled this out, it has preferred to work towards internationally acceptable codes and rules in the belief that unilateral action would undermine freedom of navigation on the high seas, and open the door to political retaliation in other international waterways.

Although freedom of the high seas is today a cardinal doctrine of international law it has not always been so. In the first part of the sixteenth century the concept was virtually non-existent. Genoa claimed the Ligurian Sea and Venice the Adriatic. The Baltic was divided between Sweden, Poland and Denmark. Portugal and Spain shared the Atlantic Ocean with the approval of the Pope, and ships sailing north of Bergen required a royal licence from Norway.

Britain herself regarded the North Sea as the 'British Seas' and required all ships to 'vail bonnet' in the presence of English ships. Vessels which failed to salute received a shot through the mainsail and then had to reimburse the British Government for the price of the shot. British claims were upheld in Admiralty Regulations until the beginning of the last century.

It was Queen Elizabeth I who first put forward the doctrine of freedom of the high seas, 'the use of the sea is common to all, neither may any title to the ocean belong to any people or private man, for as much as neither nature nor regard of the public use permitteth any possession thereof'. The doctrine was developed and refined by Grotius who based his case on two premises, namely, that the sea cannot be held or enclosed 'being itself the possessor rather than the possessed' and that 'since navigation cannot prove injurious save perhaps to the navigator himself, it is fitting that the power and right to impede this certainly should be denied to all persons'.

These arguments are clearly void in an age when a methane tanker can flatten a coastal town, and when divers can live and work underwater to a depth of 600 feet. The doctrine was always limited by the concept of territorial waters, and it has been progressively eroded ever since by fishery claims, contiguous zones, and the rights of coastal states to seabed resources. Every exception to the rule has been admitted on the grounds that the exercise of control was necessary for the existence and welfare of people living on the coast.

Right of intervention

The fact that the Torrey Canyon sank outside territorial waters did not prevent Britain taking unilateral action, and the right of coastal states to intervene in such cases has since been recognised by the Convention on Intervention on the High Seas in Cases of Oil Pollution Casualties 1969. This authorizes the State to take such measures as may be necessary to prevent or mitigate grave danger to its coastline or related interests following a maritime casualty on the high seas.

The provisions of the Convention were ratified in the Oil in Navigable Waters Act 1971 but what is needed, of course, is preventative action not remedial action. It is anomalous that countries can accept the right of a coastal state to sink a foreign vessel on the high seas when an accident threatens pollution, but at the same time assert that coastal states do not have the right to prevent such an accident by imposing certain safety standards in defined areas near their coasts.

Canada has already acted. The Arctic Waters Pollution Prevention Act asserts Canadian jurisdiction over a large part of the high seas outside territorial limits. It empowers the Government to declare any part of the waters governed by the Act to be a 'shipping safety zone' and to make regulations applicable to shipping in that area. Pollution prevention officers will be appointed with the power to board any ship within control zones, and prohibit navigation where necessary.

Pollution control zones

According to a recent survey at least sixty countries now claim a territorial sea of twelve miles or more, and Britain is in a minority in claiming only three. The Government, however, has been right to resist calls for a unilateral extension to our limits, and the assertion of national sovereignty this implies. There are less drastic ways of securing a safe seaway.

The doctrine of the Contiguous Zone allows a coastal state to prevent any infringement of its customs, fiscal, immigration or sanitary regulations which apply inside its territorial limits in an area of the sea extending beyond the maritime belt. At present this zone is restricted to twelve miles from the baseline but it is suggested that the concept could be broadened to cover all the waters overlying the continental shelf, and that within designated areas the coastal state could have jurisdiction for the purposes of pollution control along the Canadian pattern.

This solution commends itself as being an exercise of extra-territorial jurisdiction rather than an outright claim to sovereignty with all that that would entail for freedom of navigation on the high seas. It would allow countries like Britain to establish controls in danger spots like the Dover Strait, to make further provision for the safety of oil rigs and submarine pipelines, and to take any other measures for the safety of sea traffic which the situation warrants.

The International Law of the Sea Conference 1973 will provide an opportunity for proposing these reforms but if necessary Britain should act unilaterally. State practice has always been recognised as a legitimate means of developing customary international law and for elaborating new legal codes. It happened with the territorial sea. It happened with fishery limits. It happened with claims to the seafloor.

The present *laissez-faire* regime on the high seas does not strike a proper balance between the interests of flag-states in their unfettered right to navigation and the fundamental interest coastal states have in the integrity of their shores. Transport is only one among many uses of the marine environment, and in the interests of other uses and users it is to be hoped that the Government will have the foresight to act in advance of events, before a catastrophe, and the ensuing outcry, force their hand.

6. Summary of findings

MARINE POLLUTION is caused by the indiscriminate disposal of wastes or by the accidental spillage of noxious substances The first can be prevented; the second, by definition, cannot. Steps can be taken, however, to contain the number of accidents.

Voluntary pollution
The sea's capacity to degrade and dilute different substances is enormous and to exploit them as a receptacle for waste is legitimate provided that it is recognised that the self-purifying powers of the ocean are not unlimited. Unless disposals are properly regulated lasting damage may be done to other uses and resources of the sea.

There is no real evidence that waste disposal has adversely affected the life and health of the marginal seas around Britain. Where contamination persists it is highly localised, and on the few occasions when it has become widespread its effects have been shortlived. On the other hand, the volume of wastes discharged is increasing, and the shallow, semi-enclosed character of our offshore areas limits the quantity and types of waste that can be safely dumped there.

Although Britain has full jurisdiction over disposals inside the territorial sea, and over dumping from her own ships anywhere on the high seas, there is no comprehensive system of national controls. Nor is there an effective administration to enforce the rules which do exist.

Our coastal waters are put to many uses and some organisation is needed to preserve the productive capacity of this region as a total resource. It is recommended that a series of Coastal Sea Authorities should be established covering the whole of our coastline. These should be organised on a regional basis with boundaries drawn to coincide with configurations of the coast and the movement of water bodies offshore. Their jurisdiction would be confined to the territorial sea.

The new Authorities should be answerable to the Secretary of State for the Environment and they would replace the existing Sea Fisheries Committees. They should have the power to regulate all disposals and before consents are issued the effect of disposals on biological resources, recreation and public health must be assessed. The burden of proof that waste will not harm the environment should rest with the discharger.

The Coastal Sea Authorities should also have a custodial role. Before permission is given for any industrial development on the coast, or on the seabed itself, planning and licensing bodies should be required, by law, to submit the scheme to the local CSA for an environmental impact report. These reports should form an essential part of planning appeals.

Members of the CSA's should be drawn from the local authorities, river authorities, and other organisations with an interest in the maritime belt within each district. They should be financed by a precept on the rates but additional funds could be provided by charging a fee for disposals. The charge should be high enough to cover the costs of monitoring and research.

Dumping by British ships outside territorial waters is only subject to a voluntary system of controls at present. It is recommended that a compulsory system should now be adopted, with proper provision for enforcement and adequate sanctions for the violation of conditions. This follows practice overseas and is the only policy consistent with international obligations.

All dumping should be controlled, including that done by the armed services, and, again, a fee should be charged. Licences should be issued by the Department of the Environment after consultation with other ministries. The Minister should have the power to ban the dumping of specific wastes and to designate safe sites. For more difficult wastes thought should be given to setting up a Public Disposal Service.

Controls could be exercised at the loading point, and day-to-day administration of the system could be carried out by the local CSA from ports within its district. A watch on the dumping grounds could be kept by coastguards, lightships, fishery protection vessels, fishermen, and other units normally involved in policing the marine environment.

Involuntary pollution

Many noxious substances are introduced into the sea involuntarily – spilled in shipping accidents or released by mining operations on the seabed. With increasing congestion in our sea lanes and the accelerating development of our continental shelf it is unlikely that pollution from this source will decrease. This fact must be considered before giving consents for waste disposal at sea. Enough oil, for example, reaches the sea accidentally without having to cope with the deliberate discharge of oily residues from tank washings.

The Inter-Governmental Maritime Consultative Organisation is the agency for promoting safety at sea. Important work has been done but more attention should be given to offshore installations and hazardous cargoes other than oil.

68

IMCO should review the adequacy of the 500 metre safety zone around production platforms, and the safety of pipelines on the seabed.

The law of the sea does not strike a proper balance between the interests of flag-states and the interests of coastal states. The Opposition have therefore urged the Government to extend our territorial limits to the median line in the Channel so that we can exercise our own controls in these waters. Their advice should be ignored. Annexation of the Channel would undermine freedom of navigation on the high seas and invite political retaliation in other vital waterways.

A safe seaway can be secured by less drastic means. The doctrine of the Contiguous Zone allows the coastal state to prevent any infringement of its customs, fiscal, immigration or sanitary regulations, applicable inside territorial waters, in an area of the sea outside. At present the zone is restricted to twelve miles from the shore.

It is suggested that the concept could be broadened to cover all the waters overlying the continental shelf, and that within internationally designated danger spots coastal states could control navigation for the purposes of pollution abatement. This solution commends itself as being an exercise in extra-territorial jurisdiction rather than an outright claim to sovereignty with all that that would signify for the freedom of the seas.

The International Convention on Intervention on the High Seas 1969 already authorises coastal states to take measures necessary to avoid danger to their coastline following a maritime casualty outside territorial waters. What is wanted is preventative, not remedial action. The International Law of the Sea Conference 1973 will provide an opportunity to press for this reform. If necessary Britain should act unilaterally.

Further reference

Official reports

Medical Research Council: *Sewage Contamination of Bathing Beaches in England and Wales.* HMSO, 1959

Select Committee on Science and Technology: *Coastal Pollution.* HMSO, 1968

International Council for the Exploration of the Sea: *Pollution of the North Sea.* Charlottenlund, 1969

Council of Environmental Quality: *Ocean Dumping.* US Government Printing Office, 1970

Working Party on Sewage Disposal: *Taken for Granted.* HMSO, 1970

Technical Committee on the Disposal of Toxic Solid Wastes: *Disposal of Solid Toxic Wastes.* HMSO, 1970

Royal Commission on Environmental Pollution: First Report. HMSO, 1971

Joint Report of the Research Councils: *Pollution Research and the Research Councils.* 1971

Natural Environment Research Council: *The Sea Bird Wreck in the Irish Sea.* NERC Publications, 1971

Working Party on the Monitoring of Foodstuffs for Mercury and other Heavy Metals: *Survey of Mercury in Food.* HMSO, 1971

National legislation

Sea Fisheries Regulation Act, 1888
Salmon and Freshwater Fisheries Act, 1923
Public Health (Shellfish) Regulations, 1934

70

Coast Protection Act, 1949
Rivers (Prevention of Pollution) Act, 1951
Rivers (Prevention of Pollution) (Scotland) Act, 1951
Oil in Navigable Waters Act, 1955
Radioactive Substances Act, 1960
Clean Rivers (Estuaries and Tidal Waters) Act, 1960
Rivers (Prevention of Pollution) Act, 1961
Oil in Navigable Waters Act, 1963
Continental Shelf Act, 1964
Territorial Waters Order in Council, 1964
Petroleum (Production) (Continental Shelf and Territorial Sea) Regulations, 1964
Rivers (Prevention of Pollution) (Scotland) Act, 1965
Sea Fisheries Regulation Act, 1966
Oil in Navigable Waters Act, 1971
Merchant Shipping (Oil Pollution) Act, 1971
Prevention of Oil Pollution Act, 1971

International conventions

Geneva Convention on the Continental Shelf, 1958
Geneva Convention on the High Seas, 1958
Geneva Convention on the Territorial Sea, 1958
North Atlantic Fisheries Convention, 1967
Convention for the Prevention of Pollution of the Sea by Oil, 1954 (amended, 1962, 1969)
Convention on Civil Liability for Oil Pollution Damage, 1969
Convention on Intervention on the High Seas in Cases of Oil Pollution Casualties, 1969

National research centres

Fisheries Laboratory, Burnham-on-Crouch
Fisheries Radiobiological Laboratory, Lowestoft
Hydraulics Research Station, Wallingford

Hydrographic Department of the Navy, Taunton
Institute for Marine Environmental Research, Plymouth
Institute of Coastal Oceanography and Tides, Birkenhead
Marine Biological Association, Plymouth
National Institute of Oceanography, Godalming
Unit of Coastal Sedimentation, Taunton
Warren Spring Laboratory, Stevenage
Water Pollution Research Laboratory, Stevenage

International organisations

Food and Agriculture Organisation, Rome
Inter-Governmental Maritime Consultative Organisation, London
Inter-Governmental Oceanographic Commission, Paris
International Atomic Energy Agency, Vienna
International Commission for Radiological Protection, England
International Council for the Exploration of the Sea, Denmark
International Hydrographic Bureau, Monaco
World Health Organisation, Geneva
World Meteorological Organisation, Geneva